I Love Dirt!

I Love Dirt!

52 Activities to Help You and Your Kids Discover the Wonders of Nature

Jennifer Ward

Foreword by Richard Louv

Illustrations by Susie Ghahremani

TRUMPETER

BOSTON & LONDON

2008

TRUMPETER BOOKS
An imprint of Shambhala Publications, Inc.
Horticultural Hall
300 Massachusetts Avenue
Boston, Massachusetts 02115
www.shambhala.com

9 8 7 6 5 4 3

Design by DEDE CUMMINGS DESIGNS
Printed in Canada

✿ Interior printed on 100 percent postconsumer recycled paper.
∞ This edition is printed on acid-free paper that meets the
American National Standards Institute z39.48 Standard.
Distributed in the United States by Random House, Inc.,
and in Canada by Random House of Canada Ltd

Library of Congress Cataloging-in-Publication Data
Ward, Jennifer, 1963–
I love dirt!: 52 activities to help you and your kids discover
the wonders of nature/Jennifer Ward; foreword by Richard
Louv; illustration by Susie Ghahremani.—1st ed.
p.   cm.
Includes bibliographical references and index.
ISBN: 978-1-59030-535-5 (pbk.: alk. paper)
1. Outdoor recreation for children. 2. Children and the
environment. 3. Nature study—Activity programs. I. Title.
GV191.63.W36 2008
796.083—dc22
2007029744

Contents

Spring

Activities for Warm Days and Rainy Weather

Summer

Activities for Hot Days and Warm Nights

Fall

Activities for Cool Days and Cloudy Weather

Winter

Activities for Cold Days and Snowy Weather

· Contents ·

Foreword

by Richard Louv

In my great-grandparents' day, most people spent the better part of their lives outdoors. By necessity, they walked to school or work, tended the fields to grow food for the table, or fished a nearby lake. We wouldn't want to assume that generation's physical hardships, but when it came to fun, that was outside, too—swimming, boating, hiking, skiing—and the seasons were wired into their very being. They didn't have to think about nature as something to be attained, because it was integral in their lives, as it had been for untold generations. To be human was to be part of the natural world.

In the 1950s, when I was a child, I was outside playing in the nearby woods every chance I got. I fished, built tree houses, and walked with my dog, Banner, along every trail. And every year, my father drove the family on turtle patrol. My brother and I scooped up turtles crossing the country road to save them from passing cars. Some of those turtles came home with us for a season, and I learned firsthand about kindness and bootstrap conservation—as well as something about turtles. Those family outings meant a great deal to me. I passed my stories along to my two boys when they were small and called them turtle tales.

Things are very different now. Our society has largely come to regard nature as separate from normal daily life, as little more than a passing dream on the other side of the SUV's windshield. That probably won't be national park scenery, either. Visits to U.S. national parks increased at a steady clip from the 1930s until 1987, peaking at an average of 1.2 visits a person per year. But over the next sixteen years, the number of people visiting those parks dropped by 25 percent— partly because of the growing disconnect between children and nature.

As a reflection of our national move indoors, we also see lower enrollment in college undergraduate conservation programs, a drop in sales for entry-level camping gear, and fewer kids jumping on bikes. Between 2000 and 2004, sales of children's bikes fell by 21 percent, according to Bicycle Industry and Retailer News. There are even indications participation in organized sports is falling off. Meanwhile, it is no coincidence that child obesity rates are going off the charts, diabetes is a national concern, and attention deficit disorders and childhood depression continue to challenge parents and caregivers. Unfortunately, the rate of antidepressants prescribed to children has doubled in the last five years. For these and other reasons, some suggest children today may comprise the first generation since World War II to have shorter life spans than their parents.

As I wrote in my book *Last Child in the Woods: Saving Our Children from Nature-Deficit Disorder,* we are seeing children with tunneled senses and feelings of isolation and containment. What we've been ignoring as a society is nature, a potent therapy available at no cost. While nature experience should

not be seen as a panacea, study after study indicates time outdoors in a natural setting is not simply a nice-to-have activity.
It is a vital element for healthy childhood development. More
than a hundred studies of adults and children show that spending time in nature reduces stress, while other studies show that
contact with the natural world significantly reduces symptoms
of attention deficit disorder in children as young as five.

A 2005 study by the California Department of Education
found that students in schools with nature immersion programs performed 27 percent better in science testing than
kids in traditional class settings. Similarly, children who at-
tended outdoor classrooms showed substantially improved
test scores, particularly in science. Such research consistently
confirms what our great-grandparents instinctively knew to
be true, and what we know in our bones and nerves to be
right: free-play in natural settings is good for a child's mental
and physical health. The American Academy of Pediatrics
agrees, stating in 2007 that free and unstructured play is
healthy and essential for children.

Fortunately, a number of state and local governments are
beginning to catch on, and thanks to the efforts of citizen
activists and conservation groups, such as the Sierra Club,
what has been dubbed Leave No Child Inside legislation is
finding its way to governors' desks.

There is considerable interest in the subject outside the
United States, as well. Swedish researchers, for example,
focused on children in a green daycare who played outside
every day, mud or shine, and found their motor coordination
and ability to concentrate superior to that of their classroom-
bound peers. As research broadens and continues to persuade

skeptics, the public at large will respond to the evidence. I believe negative trends may yet reverse themselves. Already, a grassroots children-and-nature movement is emerging across the country, mainly at the regional level, and is growing at a quickening pace. Some of our national leaders are watching and listening and, so far, their interest is encouraging. In 2007, the U.S. Forest Service led a $1.5 million, fifteen-state effort to get more children back in contact with nature, and the Bureau of Land Management launched a "Take It Outside" initiative to bring children and families closer to their public lands.

But even if parents have the desire to get their children outside and are familiar with all the why's and should's, there will be those moments when the imagination runs dry, and then a scramble to come up with engaging things to do.

Many younger parents may not have had the opportunity to gather their own turtle tales during their childhoods, and that's why *I Love Dirt!* is especially valuable. Jennifer Ward has created a book that will serve to gently introduce parents to nature, even as parents are using it to help guide a child into the natural world. Children—and parents—learn to observe, as well as appreciate, the basic joys of getting their hands dirty and feet wet. Discoveries become shared experience. Before I saw it suggested in her book, I'd never considered gathering natural materials and attempting to construct a typical bird nest, but the idea is intriguing. I know without trying that the author is right when she writes: "Even though we can use our hands, birds have us beat in the department of nest building!"

There is no exotic setting required, no special equipment: simply flip through the book to the appropriate season—or mix and match—then find an activity (or inactivity) that speaks to you, and get out there. I say inactivity, because there are a number of wonderful suggestions that are designed to get readers to slow down, see what is around them, and pay attention to the senses. Here's one about rain that is deceptively simple, yet meaningful to anyone too much in motion to feel attached to the world: "Listen to the sound. How does it make you feel?"

When playing outside in fields or a naturally landscaped backyard, children stretch all of their senses, something they do not do in front of a screen. This book helps show how to extend our reach beyond the technological bubble. Here is the potential for much fun, many meaningful experiences—and a trove of shared stories.

Acknowledgments

My deepest appreciation to:

The folks at Shambhala Publications, especially Jennifer Brown, for her vision throughout this project.

Stefanie Von Borstel, my agent, and Lilly Ghahremani at Full Circle Literary. Thank you for planting the seed.

My family, who graciously put up with my "life in dirt" for months on end.

And special thanks to my parents, Paul and Charlene Sultan, who nurtured nature in my soul throughout my childhood. I am ever thankful and love you both very much.

Introduction

There is nothing more joyful and inspiring to watch than children discovering the world around them. Whether they're collecting fallen leaves, rolling down grassy hills, or playing in the waves at the beach—seeing that wide-eyed wonder in our children is such a gift. Regardless of the joys that nature can bring, there has been much disturbing press about the lack of quality time that children today spend outdoors, along with startling research and statistics that point to the detrimental effects this "nature deficit" can cause. And we see it ourselves. Our children rush around with too little time, too many gadgets, and too many distractions, seeking a place to play that is closest to an electrical outlet. But it doesn't have to be this way.

I Love Dirt! is a call to parents, educators, and caregivers to help recover one of the greatest joys in childhood: spending time outdoors in nature. In five minutes, you can take children outside and turn their world around. This book will help you to do that.

These fifty-two open-ended activities offer a wealth of creative ways to give the outdoors back to your children. Be it cloud racing, stargazing, making a playdate with a puddle, or building a bird feeder out of snow, *I Love Dirt!* will encourage your kids to explore, discover, giggle, exercise, wonder, and have a blast with nature. Sidebars and prompts are used throughout the book to help you explain basic science and nature concepts to children. All you need to get started is the desire to open the door to a healthy relationship with nature. Soon your little ones will also discover that spending time outdoors offers the perfect opportunity to enjoy relaxation, quiet time, reflection, and insight.

The outdoors is at your fingertips, be it a balcony, a backyard, a porch, or a playground. It is a place just waiting to be enjoyed and discovered. Time in nature is cost-free—as are the activities in this book—but the benefits will stay with your children for a lifetime. Together we can give the young minds of today—of our future—the greatest gift of all: an awakened awareness of the outdoor world.

Thank you for caring enough to pick up this book. It is the first step to opening the door of discovery just waiting for you and your children outside. Now get out there and get dirty!

Spring

Activities for Warm Days
and Rainy Weather

Spring Into Spring

It's springtime! Trees are stretching and yawning. Buds are peeking. Birds are gathering materials and building nests. Seeds are seeking sunshine and sprouting. Awaken a new sense of wonder for your children by discovering the offerings of spring.

Select a spot outdoors where you and your child can stretch and observe. Take a moment to relax and look around. Can you find signs of spring? Flowers? Buds on a tree? Seedlings? Observe the earth around you as it prepares for a season of

Help Me Understand

Q: What makes new plants sprout in the spring?
A: They get more sunlight than they were getting in the winter. In the spring, the days get longer, brighter, and warmer. These things help new plants to grow.

new growth. How many different signs of spring can you count?

Go on a spring hunt, seeking out new growth in your region. Landscapes tend to transform as if overnight when spring beckons. Search for signs on tree branches, flower stems, and succulents such as cacti. Search for animals that have returned or awakened from a winter slumber, now that spring has sprung.

Spring is soft: new grass, new buds, gentle breezes, gentle sun. Experience the softness of the season. Feel the breeze on your faces. Together, take your shoes off and gently brush the soles of your feet over the new blades of grass. Gently glide the palms of your hands over them too. Let spring's softness tickle your senses.

Crouch down with your child and practice sprouting like blades of grass. Slowly straighten your legs and grow upward. Extend your arms toward the sky, breathing, relaxing, and taking in spring. Crouch low and pretend that you're a flower, getting ready to bloom. Then do it—slowly. Get into action and find a smooth, grassy area where you can tumble in somersaults.

☑ *Stimulates relaxation and exploration*

2

Bouquet of Color

Have you ever taken the time to look at the colors nature offers? We often think of green as the color synonymous with nature in regard to trees and grasses. Blues come to mind when we envision the sky, oceans, streams, and lakes. What other colors are out there? Mother Nature promises a rainbow and then some. Take a walk with your child around your home, in a park, or even along a sidewalk and seek out the colors of nature.

How many different colors can you find in flowers? Which color flower is the most common? Yellow? Blue? White?

Do the flowers you can find have visitors, such as butterflies, hummingbirds, or bees? These visitors are often pollinators. A flower's color will attract them. The color seems to shout "I'm over here!" to bees and butterflies. Pollinators are very important to plants. Without them, plants would not grow.

Take note of petal shapes. Are all flowers shaped the same?

Help Me Understand

Q: Why are there so many colors in nature?

A: Colors serve a purpose. They might camouflage an animal to protect it from predators, animals that hunt other animals. A bright color may help attract an animal or a bug, just as a flower attracts bees, which can help it to grow. Some colors give warning, such as "Watch out! I'm poisonous."

How many different types of petals can you find as you look at different flowers?

Do all flowers have the same number of petals? Search for different types of flowers and gently count the number of petals each has.

What other colors can you find in nature? Look high and low. Seek out the colors of the rainbow in order: red, orange, yellow, green, blue, and purple.

What colors can you find in living things in nature? What colors can you find in nonliving things in nature?

☑ *Stimulates awareness of one's surroundings and concentration*

3

Move Over, Clover

Young children learn basic concepts in primary school, such as using their senses, identifying shapes, and identifying colors. Spending time outdoors can be educational and inspirational, making these important concepts more meaningful.

Take a walk outside with your child. It could be in your backyard, in a park, along a beach, or even along a sidewalk. Explain that you're on a treasure hunt for the color green. To add an element of fun to the treasure hunt, don a green shirt or socks, or if you're near a beach, a green cap or flip-flops.

Help Me Understand

Q: Why are so many plants green?
A: The green color found in leaves helps plants get energy from the sun. All living things need energy. (We eat food to get energy, right?) Plants use the energy they get from the sun to make oxygen for us.

Display and share the green items you discover in nature, while encouraging your child to do the same.

Search for as many different shades as you can find: deep green, pale green, bright green, dull green. Count the number of different hues you find. Ask your child to select a favorite shade of green.

Take note of where green is found. Practice counting the number of places you and your child locate green, such as stems, leaves, grass, mosses, stones, rocks, water plants, or shells. Look closely at places you wouldn't ordinarily notice, such as the trunks of trees or in sidewalk cracks. You might be surprised by what you discover.

If your locale is appropriate, challenge your child to search through a patch of clover for a four-leaf clover. Demonstrate how to use your fingers to separate the clover, closely observing each plant for four leaves. If a certain patch of clover doesn't produce results, consider regrouping in a different patch. But don't give up the search! They're out there.

☑ *Stimulates observation skills and awareness of the environment*

4

Hear, Here!

Children learn about their five senses through experiences. Nature offers many ways to explore our senses—from a soft breeze on our face to the bright red of a flower to the taste of a freshly picked berry. This activity focuses on the rich sounds of nature. Indoor environments are often polluted with a lot of noises, such as televisions and the daily hum of electrical appliances, that are so ingrained in our brains that we don't even notice them. Encourage your child to take in the pureness of natural sounds by focusing on them.

First, locate a nice spot where you can spread out a blanket or towel and take a seat with your child. The deeper into nature you can go, the more natural sounds you'll hear; but wherever you are, I assure you that you'll be able to focus on something. This quiet activity may provide some good snuggle time—feel free to bring along a favorite stuffed animal.

Sit and speak quietly, encouraging your child to listen.

Help Me Understand

Q: Are there any animals that don't have ears?
A: Some animals, like fish, have ears inside their heads instead of on the outside.

Explain that you're going to create silence, and see what it will bring.

Does silence have a sound? Shhhh. Listen. What do you hear?

Do you hear birds busy in the trees and celebrating spring? How do they sound? Happy? Argumentative? Joyful? Chatty? How many different bird sounds do you hear? Can you see any of the birds?

What else do you hear? Filter through the everyday sounds, and focus only on the sounds of nature. Can you hear wind? Insects? Close your eyes and continue listening to the sounds that silence brings. What sounds of spring can you hear? Water dripping? A breeze whispering through tree branches?

End your quiet time listening to nature by whispering something sweet in your child's ear.

Everyone Has Ears

Make a point of noting the many different types of ears that animals have. Rabbits have tall ears. Birds have tiny holes for ears. Bats often have wide ears. Dogs have lots of different types of ears. Some ears move forward and back. Some remain stationary. Animals in the wild often use their ears to warn them of danger and to communicate with one another.

☑ *Stimulates awareness to one's surroundings, relaxation, and appreciation of the senses*

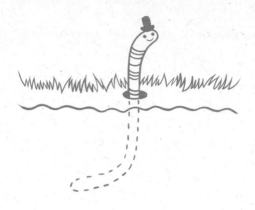

5

Wiggly for Worms

There are so many creatures out there to explore. One of the most bizarre yet ordinary is the worm. Earthworms often surface when the ground is moist. Scientists aren't exactly sure why earthworms surface following a rain. Different theories suggest that the soil might be too moist, making it difficult for them to breathe; that the newly moistened soil makes it easier for them to move through to the surface in order to burrow in new areas; or that they surface to look for mates. Earthworms prefer moist, dark climates and will dry out and die if they're aboveground on a sunny day.

Help Me Understand

Q: What are worms?

A: Worms are animals. They are invertebrates, which means they do not have a backbone like you or I do. Worms breathe through their skin. They do not have eyes, but they are sensitive to light and dark and vibrations. They burrow through the soil, eating and digesting dirt along with organic matter, such as old leaves. This helps keep the soil healthy and rich so that plants can grow.

Search for worms following a rainfall, and watch a worm in action. How does it move? Where is it going? What is it doing? A worm in action is a worm helping the soil to be a healthy place for plants to grow.

Wiggle like a worm. Stretch as long and thin as you can. Shrink as short and stubby as you can. Wiggle! A worm may not have bones, but it has many muscles that allow it to stretch and shrink and move around.

While you're outside, consider the wet ground that earthworms love. How does the earth feel beneath your feet when it's wet? How does it sound? Does it feel different to walk on moist ground than dry ground?

Amazing Earthworms

Earthworms are vital to the health of soil and plants. They burrow through the ground, creating airways for water drainage and oxygenation of the soil. They eat organic matter, such as fallen leaves, which after it's digested becomes part of the soil we need to keep plant life growing and healthy. They fertilize the soil and help make it rich.

☑ *Stimulates wonder, curiosity, and appreciation for living things*

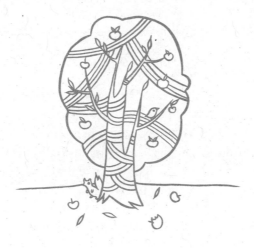

6

See That Tree?

Have you ever looked at a tree? I mean, truly looked closely at a tree? Probably not. Trees are one of those magnificent things in nature that we too often take for granted. Yet trees sustain us by breathing out what we breathe in, while also providing wood for homes, bridges, buildings, boats, furniture, art, accessories, clothing, paper, and so on.

For humans, trees also provide food, shelter, shade, and so much more. Do animals rely on trees as much as we do? How

Help Me Understand

Q: How old are trees?

A: Trees begin their lives as seeds and take many years to grow. It can be difficult to measure the age of a living tree, unless you know exactly when it was planted. Scientists often study the rings that form within a tree's trunk to measure its age. They count the number of rings, which help determine its age.

do they depend on trees? Are trees important to the land around them? Select a tree and look at it. Reflect with your child.

What does that tree offer to living things—shade, shelter, food, transportation? What creatures might use it? Make a verbal list and see how many nature dwellers you can come up with. Then sort your list by categories such as the following:

- Animals that live in or on a tree
- Animals that get their food from trees
- Animals that build with trees
- Animals that travel among trees

Or looking at it from a different angle, consider which animals might depend on specific parts of a tree:

- Which animals use a tree's branches? How (such as birds for nests, squirrels for traveling)?

• Which animals use a tree trunk? How (such as, wood-peckers for finding food and building nests, insects for traveling and nesting)?
• Which animals use a tree's leaves? How (such as, birds for shade, bugs for food and shelter, squirrels for nest-building materials)?

Think about all the ways trees differ from one another. Consider their height and width, their bark, their leaves, their branches. How many different types of trees can you find in your area? Look around and notice their shapes and forms. How are they alike? How are they different? Do all trees look the same? Are they all the same size?

Try this: Stand like a tree. Sway like a tree. Hug a tree! Get to know the trees in your area and appreciate them.

☑ *Stimulates awareness, appreciation, and understanding of plant life*

7

Leaf Looking

Time to go on a leaf hunt! Get your budding biologist geared up for some leafy fun while you explore the types of leaves that nature produces. You won't be disappointed.

Nature provides plant colors and shapes for a purpose. It's easy to ooh and aah over these colors without taking the time to consider the purpose each actually serves. The pale hues of plants in the desert absorb less sunlight. Since the desert is an environment with an abundance of sun, desert plants have adapted by color, shape, and form to survive in that particular environment. Desert plants also have a waxy coating to help store precious water and thorns that warn plant-eating animals to stay away. At the other end of the spectrum, plants found in moist and cloudy climates are darker in color, as dark colors absorb more sunlight than light colors.

The shape of each leaf and petal serves a special purpose as well. The leaves on some plants are designed to absorb the maximum amount of sun that a plant needs to survive. Some

Help Me Understand

Q: What are leaves?
A: Leaves work as an essential organ in plants, just as the heart is an essential organ in humans. The leaf is the primary place on a plant where photosynthesis takes place. Photosynthesis is a complicated process: the leaf uses sunlight, water, and carbon dioxide and creates a sugar. The final process results in the plant emitting oxygen as part of the air we breathe.

leaves are designed to catch rainwater to feed the plant. Some leaves even funnel water toward the plant's roots. Rainy environments produce plants with very large leaves. Dry environments produce plants with very tiny leaves.

Explore your area and discover what type of leaves you can find. Take note of size; shape; color (light, dark, or somewhere in between); and texture (smooth, waxy, fuzzy, or bumpy).

There are two primary types of leaves: netted and parallel. Netted leaves have branched veins, such as the leaves found on an oak tree. Parallel leaves have veins that run in the same direction, such as those found on a corn plant or tulip. Look for leaves that fall into both categories. Which type can you find the most of?

As you search for and observe leaves, can you find evidence of animal life among them? Do you notice nibbles? Droppings? Nests? How might an animal use a leaf?

Collect different tree leaves and press them between the pages of a large book. Sort and classify your tree leaf collection by size, shape, and/or color. Borrow a tree book from your local library and see how many different types of leaves you can identify as you spend time outdoors searching and collecting leaves. How are tree leaves different from the leaves found on bushes?

Note: Some bushes, such as poison ivy and poison oak, can create rashes on skin. Be mindful of which bushes or vines you allow your children to touch.

☑ *Stimulates awareness and appreciation of plant life*

8

Going On a Bird Hunt

Bird-watching is great fun and such a simple thing to do. What's more, birds can be found almost everywhere, from city streets to rolling fields. Big birds, little birds, bright birds, dull birds—they're all around us. Connecting with the bird world is as simple as stepping outside and opening your eyes and ears to their amazing sights and sounds.

Scoot your chicks to a park, backyard, canyon, beach, forest, or lake and take their ears and eyes to the skies. Begin a scavenger hunt for birds. Use this checklist and motivate your children to use their eyes to try and spy the following:

- A brown bird
- A colorful bird
- A large bird
- A tiny bird
- A flying bird
- A bird on the ground
- A bird's nest
- A bird in a tree or bush
- A bird with long tail feathers
- A bird's track on the ground
- A bird that is singing or chattering
- A feather
- A bird with spots
- A bird with stripes on it (above the eyes, along the wings, or on its chest, for example)
- A bird hunting for food or eating

When you're done, talk about your results. What was the easiest bird or item on the list to find? Which bird or item was

Help Me Understand

Q: What makes a bird a bird?
A: Even though birds are warm-blooded, just like mammals, they have characteristics that set them apart. All birds have wings, feathers, two legs, hollow bones, and beaks, and they all lay hard-shell eggs.

the most difficult to find? Ask your children to think about why some were easier to locate than others. Were some of the birds common ones that you might see all year long? Were some of the birds species that you see less frequently?

Attracting Birds

Encourage the little chicks in your life to attract feathered friends to your own backyard. Chances are some birds live in your area year-round, while others visit seasonally. Attracting birds to your backyard is simple, regardless of whether it consists of a small patio, a porch, or rolling acres. Here are some ideas:

- Hang feeders or spread out food to attract some birds. They love raisins, peanuts, sunflower seeds, peanut butter, and orange wedges.
- If you and your young brood feel adventurous, you might consider creating a habitat for a certain type of bird. A habitat consists of food and shelter appropriate for a particular species, a place to raise its young, and water.
- Place a birdbath or water dish out for their use. Birds love water. A shallow ceramic or clay water dish intended for flowerpots works well. Place a rock in the center of the dish so the birds have something to land on. Be sure to change the water regularly.
- If your backyard consists of a porch or patio area, consider hanging a hummingbird feeder, since larger seed- and fruit-eating birds can be quite messy. Buy a hummingbird feeder (see if there's a bird store in your area or check

your local hardware store) or make one out of recyclables in your home. Mix their nectar by combining one part sugar and four parts water. Do not use food coloring in the mixture! Instead, simply tie a small piece of red ribbon to the feeder to help attract hummingbirds to it.

✓ *Stimulates curiosity and appreciation for bird life*

9

Be Like Audubon

John James Audubon (1785–1851) was an artist who spent much of his life observing and drawing birds. He is also recognized as the first person to experiment with bird banding; he captured eastern phoebes and tied pieces of string around their legs, leading to the discovery that these birds returned to the same nesting spots each year.

Not only was Audubon an ornithologist—a scientist who studies birds—he was also an artist. Allow your child to work like an artist and scientist as she studies the birds in your backyard and beyond. Encourage her to become a "bird brain."

Study and observe the birds around your yard and neighborhood. Encourage your child to keep a journal and document and draw the different types of birds she sees.

- What sizes and shapes are the birds?
- What colors are they?
- Do all birds move in the same way?

Help Me Understand

Q: How many different types of birds are there?
A: There are about nine thousand different bird species in the world.

Pay close attention to beaks. Do all bird beaks look the same? There are different types of beaks and each serves a specific purpose. For example, short beaks are seed beaks, perfect for cracking nuts and seeds open. Long and narrow beaks may be used to sip nectar, whereas sharp and pointy beaks may be used for tapping trees and eating bugs. Hooked beaks are used for hunting animals, such as mice and rabbits.

Make note of all the different types of beaks birds have. What might a bird use its beak for, other than eating? Think about this: birds don't have hands. And while we're on the subject of extremities, encourage your little birder to observe the different types of feet birds have. Do birds that live near water have the same type of feet as birds who live in dry areas?

A Bird Brain's Library

Bird books are a fantastic tool to keep on hand for your budding ornithologist. Check your local library, thrift shops, or bookstore for bird guides that your child can refer to. Some of my favorite resources include the following:

- *The Audubon Society Guide to Attracting Birds*
- *National Audubon Society First Field Guides: Birds*

☑ *Stimulates keen observation skills, relaxation, and curiosity*

10

A Little Birdie Told Me

Birds make up part of nature's music. When various birdsongs and birdcalls are combined with wind whistling through trees, water babbling in a brook, and the hum of insects, you've got quite a symphony of music available to your ears. Turn off the iPod, the television, the computer, and the stereo. Step outdoors to give your child a chance to hear music that is pure.

Most birds make two types of sound: songs and calls. Birdsong is normally created by male birds in order to attract a

Help Me Understand

Q: Do birds have feelings?
A: Scientists study animal behavior. Ornithologists (scientists who study birds) say that birds do indeed experience emotions, such as fear and excitement. Whether or not birds feel happiness is still being studied, discussed, and questioned. What do you think?

mate or defend a territory. Birdcalls are used by both sexes as a form of communication.

Encourage your child to spend time outdoors listening for the presence of birds. Make note of the following:

• When birds are most vocal (early morning, afternoon, evening, or night)
• When birds are least vocal
• How songs and calls vary by time of day
• How songs and calls vary by season
• How songs and calls vary by a bird's action (such as feeding, nesting, or perching)
• The number of different birdsongs you can document
• The number of different birdcalls you can document
• Which type of song or call you hear most often
• Two birds calling back and forth to each other

Encourage your child to imagine or predict what each type of call or song is saying. Target one specific bird sound and listen. Does it sound joyful? Angry? Relaxed? Tense or fearful? Ask your child what might make a bird feel fear (such as a threat to its territory or a sense of danger) or what might make it feel excited (such as locating a new source of food, eggs hatching, or finding a mate).

After listening to birdcalls and becoming familiar with them, have a birdcall contest!

☑ *Stimulates a sense of discovery and appreciation for other living things*

11

Up, Up, and Away

One reason we find birds so fascinating is their ability to fly.
We all wish we could soar through the sky like a bird. With
their hollow bones, dynamic bodies, feathers, and wings, birds
are the masters of the sky—the masters of flight.

Select a spot where you and your child can sit outdoors and
quietly bird-watch. Study birds flying, taking note of things
like these:

- Where birds are flying (Are there birds soaring high up in
  the clouds? Are there birds flying among tree branches?
  Are there birds taking long flights across the sky or bod-
  ies of water? Short flights from one spot to another?)
- How precise birds are when they land
- A bird that is flapping its wings in flight
- A bird that is soaring, without flapping its wings
- Birds that are flying in a group
- A bird that is flying by itself

Help Me Understand

Q: Do all birds fly?

A: Not all birds take to the sky. We can't forget penguins, who have mastered flying through the sea. Other species of birds that can't fly in the sky include the ostrich, emu, cassowary, kiwi, kakapo, and rhea.

After observing a variety of birds in flight, ask your child to imagine that he is a bird. What type of bird would he like to be, and why? Ask him to close his eyes and imagine that he's a bird flying on a journey. Where is he going? What does he see? How does it feel to fly? What is his favorite type of bird food?

Sit quietly and simply let your imagination soar on wings. Share your own bird journey with your child, answering the same questions you asked him.

Encourage your child to act like a bird. Flap your arms like wings. Use your feet to hop, hop, hop across the ground or waddle like a penguin.

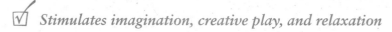 *Stimulates imagination, creative play, and relaxation*

12

Build That Nest

Birds are amazing architects, and it's always a treat to see the results of their handiwork in the form of nests. In the fall and winter, they are easier to see among a tree's bare branches and limbs. During the spring and summer, you might locate a nest based on bird chatter coming from a specific source (hungry chicks at lunchtime). When you do locate a nest, be certain to observe it quietly. Nesting birds need privacy and

Help Me Understand

Q: How do birds build nests?
A: They use their beaks to gather and collect materials. They use their beaks, feet, and even their bellies to create the actual nest, which might take a few days or a few weeks to build, depending on the species.

space. After all, how would you like it if you woke up in your bed to find a group of strange-looking animals staring at you?

Once you've found a nest or two or three, take note of their shape. Is each one shaped the same? Can you tell what materials were used to build them? Are they the same size?

Consider their location. Are some located higher than others? Are some easier to spot than others? Discuss the findings with your child.

Birds build the largest variety of homes of any wild animal species. Some nests are cup shaped, some are woven, some are baskets, and some are domes. Some hang, some are cemented to the sides of walls and under eaves with mud and spit, and others are simply constructed with loose sticks. Some exist on a simple ledge or in a shallow depression in the ground. A tiny hummingbird's nest is constructed with spiderwebs, which allows the nest to stretch as the chicks grow bigger.

Just how do birds build such incredible nests? After all, when you think about it, they don't have hands. Try to test your own bird's nest-making skills. Collect items in nature (such as grasses, leaves, twigs, fiber, and mud) that a bird might use to build a home. Try building a bird's nest with these materials. How difficult is it? Even though we can use our hands, birds have us beat in the department of nest building!

✓ *Stimulates observation skills, appreciation, and a sensitivity toward living things*

13

Rain, Rain, Come Again

Who says rainy days are to be spent indoors? Next time the clouds decide to sprinkle water down on your home, dash outdoors and play in the sky, clouds, and rain.

Spend time under an umbrella with your child, and listen to the rain pitter patter against it. Watch the rain as it rolls and drips from your umbrella to the ground.

Listen to the sound. How does it make you feel?

Catch the rain in your hand and let it pool there. Does it absorb into your skin or roll off?

Let the rain fall on your face with your eyes closed. How does it feel?

Explore just where the rain goes and how it reacts to the surfaces it meets. Are puddles forming? Where and why? Where does the rain disappear or absorb into surfaces? Where does it gather and stream?

Are there any animals or bugs out enjoying the rain? If so,

Help Me Understand

Q: Why does it rain?
A: Rain is created when moist air rises. As it rises, it cools. The water in the air clings together (condenses) or forms into water droplets, creating clouds. If enough moisture condenses together, the water falls as rain.

which types can you find and observe? Can you hear animals or bugs calling through the sound of the rain?

If the rain shower is brief and the sun begins to peek out, can you spy a rainbow?

When you are done enjoying the rain, be sure to go back indoors and dry off.

☑ *Stimulates a sense of adventure*

14

Puddle Jumpers

Splishy, splashy water. Wonderful rain. Earth is called the blue planet, because water is its main life force. Amazingly, only 3 percent of the water on Earth is fresh water, with the remaining 97 percent belonging to oceans. Of the fresh water, almost 70 percent of it is contained in glaciers and ice caps. Kind of makes you look at puddles with a bit more regard and appreciation than usual, yes?

Plan a puddly day with your child following a rain, so you can explore the wonders of puddles in person.

Find a clean water puddle. Talk about it with your child. What made it? How deep is it? Why is it here?

Locate a stick to use as a measuring tool. Estimate together how deep you think the puddle will be by marking a spot on the stick. Insert the stick into the puddle to see how deep it actually is. Was your estimate close?

If the puddle appears clean, stomp and jump in it, shoes and all. Never jump in with bare feet—there might be a sharp object hiding under the water's surface. Make a splash! Jump, jump, jump in your newfound puddle.

After splashing in the puddle, think about it some more. Ask your child if she thinks the depth is the same as it was before she jumped in it. Why or why not? Time to reestimate the puddle's depth. Use your stick and measure the depth again. Did it change?

Now locate natural objects such as pebbles, stones, nuts, pinecones, leaves, or flowers. Toss them into the puddle. Do all of the items make a splash? Which make the biggest splash? The smallest splash?

Look at the surface of the puddle. Does it act as a mirror? What does it reflect? What happens to the surface when an object touches it? How long does it take for the surface to become smooth after it has been disturbed?

Collect more items found in nature. Ask your child to predict which objects will float and which objects will sink when placed in the puddle. Sort your collected items into two piles: a sink pile and a float pile. Place the items in the puddle. Observe which ones sink and which ones float. How many of

Help Me Understand

Q: How are puddles made?

A: Puddles are usually made from rain, which collects in depressions (low places) in the ground. Depressions can be found in lots of places in nature, including on large rocks, in dirt, in low areas of lawn, on roadsides, and along shorelines.

your estimates were correct? Can you make the items that float also move across the puddle? Try blowing on or fanning the puddle. Toss a splashy object in. Did your floating object move?

Revisit your puddle the next day or every few days, if possible. Has it changed in size or depth?

☑ *Stimulates wonder, experimentation, and a feeling of exhilaration*

Summer

Activities for Hot Days
and Warm Nights

15

Digging Dirt

Time to get dirty! Dirt can be the most delightful play item in nature. Here's the scoop: it washes away from hair, clothes, skin, and toenails, so fret not about allowing your child to get a bit messy when outdoors. If you're concerned about stains, try to designate play clothes that are allowed to get dirty. (And keep in mind that a few stains are a small price to pay for the benefits of imagination and creative play outdoors in the fresh air.) The important thing is to encourage your child to play freely, without fear of getting dirty.

In a patch of dirt, provide your child with small shovels, cups, and old spoons. Objects in nature work well for dirt play too—sticks, stones, leaves, nuts, flower petals, and bark.

What can you build up with dirt? Mountains? Volcanoes? Walls? Castles? Animal shapes?

What can you channel when you tunnel with dirt? Scoop away earth and add water to the mix to create moats, miniponds, and a winding river. Watch how the water reacts with your creations. Pour water into a channel or area you've dug away and watch it pool. Does the water remain stationary? Does it absorb? Does it flow from one area to another?

Pour water onto a mountain you've made. What happens to its shape? As the mountain crumbles, it is eroding, just as real mountains erode in weather over time.

Create buildings or walls, and decorate them with natural items, such as twigs and pebbles or rocks. Allow your child to dig playing with dirt.

Mud!

Let's not forget dirt when it's wet. Mud's squishy texture can make it fun to explore. Squish mud through your fingers and

Help Me Understand

Q: What is dirt?
A: Dirt is a mixture of all kinds of things: broken rock and stones, minerals, and organic matter such as broken-down bits of plants.

in your hands. Can you shape it? Can you form it into bowls or cups? Can you roll it into balls?

Press objects such as leaves or twigs into mud. What types of impressions do they make? Walk through mud and let your feet squish in it. Did you leave tracks?

Can you paint or draw with mud? Use your fingers and hands to create mud pictures.

☑ *Stimulates creative play and problem solving*

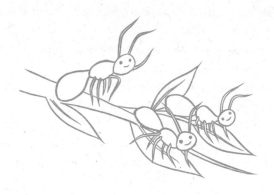

16

Antics

Probably one of the easiest insects to watch in action is the ant, and what antics they perform as they go about their very busy workdays. Ants can be found in practically any outdoor environment—from sidewalks to playgrounds, from a small patch of grass to a large field.

Scientists who study bugs are called *entomologists*. A true bug is an animal that has a mouth part that can pierce and suck, like a mosquito or ladybug. Ants are not true bugs, even

Help Me Understand

Q: How do ants smell? Do they have noses?
A: Ants use their antennae—the two long, thin body parts on their heads—to smell with.

though they have mouths and are in constant search for food. They are part of the insect family. Insects have three main body parts: a head, a thorax, and an abdomen. Insects also have six legs.

Take your budding entomologist outside and locate an anthill. (Warning: All ants can bite, and their bites can hurt! Do not hold or handle ants of any size.) Spend time watching them in action. Follow their trail and see where it leads you.

Ants follow a trail because they are searching for food. When one ant finds food, it leaves a scent trail for the other ants in its colony to follow, telling them where they can find the food supply. When you see ants following a trail, it means they are off to get provisions for the colony.

Try this fun experiment to watch how ants communicate with one another by leaving a scent trail. Collect several small twigs and place them end to end to create an enclosed space not too far from an anthill. Don't create a high enclosure; make it flat and wide. Drop some sugar or cracker crumbs within the enclosed space.

Wait for the ants to discover your gift. Soon they will find the food you've left for them, and as they take it away, they will leave a scent trail so they can return for more. Other ants in the colony will quickly catch the trail's scent and follow it to the food source too. Once you have a trail of ants in pursuit of the food, carefully remove the sticks. Observe what happens: the ants become confused because their scent trail has disappeared.

☑ *Stimulates discovery, exploration, and understanding of living things*

17

Sun Fun

Sun seekers, unite! Pick a sunny day to go outdoors and explore just what in nature seeks out the sun. For people, sun provides essential vitamins important to our health. (However, too much sun can be harmful, so remember to wear sunscreen and a hat and to drink water if you're going to be out in it for very long.)

Explore with your child outdoors. First, find a spot where you can sit, relax, and simply observe nature in action. Ask your child these questions:

- What animals are active during the day?
- Are there bugs out and about? Birds? Bees?
- Which animals like the sunshine?
- Can you spy a reptile?

Reptiles, which include lizards, snakes, turtles, tortoises, crocodiles, and alligators, are cold-blooded. This means they

Help Me Understand

Q: What is the sun?

A: The sun is a star. Stars are giant balls of hot gas. Our sun is the closest star to Earth. It is the perfect distance away—not too close to be too hot, and not too far away so that we can't enjoy the energy it provides.

cannot regulate their body temperature and often seek the sun to warm themselves. Walk around and look in sunny spots, such as on rocks or walls, to see if you can find a lizard. If you're near a pond or stream, you might see turtles sunning themselves on a log.

Notice the plants growing around you. Which direction are they facing? If there is a tree nearby, compare one side of its trunk to the other side. North-facing trunks, which get less sunshine, will look different from south-facing trunks, which get more.

If you are able, compare the north and south faces of hillsides or small mountains. How do they look different?

The amount of sun a place receives affects what animals live there and how plants grow there.

Go on a treasure hunt and search for sunseekers. How many things in nature can you find enjoying the sunshine? Don't forget to count yourself!

☑ *Stimulates observation skills, exploration, and curiosity*

18

Butterfly Café

Butterflies are one of nature's most delightful insects. With their delicate manner and aesthetic beauty, it's fun to watch them in action. Butterflies are vital to the health of gardens and plant life. Each time they visit a flower to drink its nectar, they help to pollinate it. Pollination creates new seeds that can grow into new plants. (Pollination happens when pollen from a flower reaches another part of the flower called a pistil.) Butterflies and other animals, such as bees and bats, help pollen travel from flower to flower, ensuring pollination, which ensures new seeds to make new plants. Without pollinators, plants cannot survive. Different types of butterflies rely on different flowers for food, but certain flowers can provide food for a wide variety of butterfly species.

Find a garden where flowers are in bloom and spend time with your child observing butterflies. Look for these behaviors:

- Feeding. A butterfly uses its proboscis, which acts like a straw, when drinking from flowers. If you watch very carefully, you can see this process.
- Sunning. Butterflies must maintain a certain body temperature before they can fly. If a day is chilly, you might see a butterfly soaking up heat to warm itself.
- Flying. Butterflies are excellent flyers. Certain species, such as the monarch, migrate thousands of miles each year. Without flowers to sustain them, they could not complete their migration.

Creating Your Own Garden

Creating a butterfly garden is a fun and simple way to bring butterflies to your own backyard, where you and your child can spend hours delighting in their behavior. You don't need a large spot of land, and flowerpots will do if you don't have access to a backyard. Butterflies rely on specific plants, called *host plants,* for laying their eggs. Providing host plants for butterflies may offer the opportunity to watch caterpillars as well, because these are the plants a caterpillar will feed on after hatching from a butterfly's egg.

Help Me Understand

Q: What is a butterfly?
A: A butterfly is an insect. All insects have three body parts and six legs.

Check with your local nursery to get the names of native nectar flowers that grow in your region. Planting native flowers is a wise choice for two reasons. First, native flowers require less maintenance, as they do just fine growing on their own in your region. Second, native flowers will sustain the wildlife in your area, such as migratory butterflies, pollinating animals, and even hummingbirds.

Once butterflies frequent your garden, take note of their shapes, colors, and sizes. How many different species can you spot? Start a butterfly journal to keep track.

☑ *Stimulates stewardship, respect, and sensitivity for animal life*

19

Ladybug, Ladybug

Ladybugs are a garden's best friends. They help keep it healthy by eating pesky bugs, such as aphids, that can be harmful to plants.

Most children are familiar with ladybugs, although they may be more apt to recognize a ladybug from a cartoon or manmade designs seen on clothing, toys, and furniture. Get your kids outdoors and into a garden in search of real ladybugs! They're hard to miss, with their bright red wing coverings and black polka dots. The bright red color serves as a

Help Me Understand

Q: Are all ladybugs ladies?
A: No. Ladybugs can be boys or girls. Even boy ladybugs are called ladybugs.

warning to birds and other animals looking for food. Ladybugs are not poisonous, but they may leave a very bad taste in the mouth of an animal that eats them.

As you search for ladybugs, check out these things:

- Do all ladybugs look the same? How are they alike? How are they different?
- Do all ladybugs have the same number of dots?
- How does a ladybug move from one place to another?

The ladybug uses its antennae to seek out meals, such as aphids. When you find a ladybug, handle it gently, and always place it back where you found it.

 *Stimulates appreciation and respect for insect life*

20

Handy Plants

Spring and summer make a busy season for plants, with all their sprouting and flowering, followed by pollination and seeding (for those lucky enough to be pollinated). Seeds have amazing ways of traveling, from floating through the air (think of dandelions) to attaching themselves to a squirrel's tail. It's important for some seeds to travel away from the plants that they came from in order to find a spot of soil all their own. Late spring or summer is an excellent time to take your little sprouts out and about for a seed search.

Get an old, itchy sock that no one likes to wear or that is missing a mate (an old, knit winter glove that you can part with will work too). Have your child place her hand inside the sock or wear the glove. Then take a walk together. As you pass grassy areas, have your child run her hand over the grass tops. Flower patches that are past the blooming stage are great things to brush over too. Experiment with a variety of

Help Me Understand

Q: Where do seeds come from?
A: Seeds come in all shapes and sizes. Flowers produce seeds if they are pollinated. Pollination occurs when pollen from a flower reaches another part of the flower called the pistil. Seeds are made in the pistil.

vegetation. Soon you'll see you have tagalong visitors (seeds) clinging to the knitwear.

Once you've collected some stowaway seeds, take the sock home. Place some soil inside it, then place it on a tray in a sunny window. Water your sock and wait to see what grows!

Simple Windowsill Gardens

Another easy way to help seeds sprout is by collecting a Styrofoam egg carton or an empty milk or juice carton (which you can cut in half). Fill the empty area with soil, collect some seeds, place them where they'll get some sun, and watch what grows.

☑ *Stimulates wonder, physical activity, and appreciation for nature*

21

Wow, Water!

If you ask most children today to think about water, they will visualize a faucet, a sink, a bathtub, or a shower. Water is simply something that's turned off and on. How sad that they fail to realize that water is a part of nature, not just a product from a spout. It's time to get outdoors and explore the wonders of water, a main life source for every living thing on Earth.

The water we enjoy and need rises from the ocean, gathers

Help Me Understand

Q: What is water?
A: Water is a clear liquid that comes from rain, snow, sleet, and hail.

as clouds, and falls as rain in a cycle. We can't live without it. Provide your child with an opportunity to learn about water and its properties.

Find a spot outdoors that is suitable for some water play. Perhaps a nearby shore or stream, a fresh puddle, or a patch of early-morning grass with dew clinging to it. Watch what water does.

Have your child find two dry leaves—fresh is okay, but not wet. Two dry blades of grass will work as well. Ask him to press the two leaves together. What happens? (Nothing.)

Now wet the two leaves by rubbing them in dewy grass or putting water on them. Have your child press them together again. What happens? (The leaves stick together.)

Explain to him that water is *cohesive*. This means that water likes to stick to water. Because both leaves were wet, the cohesive properties of water made them stick together.

Try to locate other objects in nature that will stick together when wet, such as feathers, grass, grains of sand, dirt, tiny shells, and so on.

Will different types of objects stick to one another? Try sticking a leaf to a blade of grass.

What natural objects won't stick together, even when wet (rocks, twigs, medium to large shells)?

Can you think of ways water's sticky properties work on your body (think about your hair when it's wet compared to when it's dry)?

☑ *Stimulates exploration and awareness of water and its properties*

22

Water-Hole Watching

Water sources offer a wealth of observation opportunities. Wildlife depends on natural water sources to bathe in and drink from. Humans have water sources such as faucets and fountains, sinks and tubs. In contrast, wild animals depend on Mother Nature.

Observing what species depend on water in nature is simple to do. From puddles to oceans, chances are something wild makes its home in, on, or near the water. Here are some things to take note of:

Puddles

- Watch for birds that will stop by for a drink or a quick bath. Some insects, such as bees and butterflies, may drink from puddles as well.

Streams, Marshes, Ponds, and Creeks

- Search for evidence of wildlife along the banks by locating animal tracks in the mud.
- Look for evidence of beavers in the form of chewed logs and branches.
- Listen. What sounds do you hear—frogs croaking, birds, bugs?
- What's living in the water? Can you see fish, bugs, turtles, tadpoles?
- Some ponds are temporary, only lasting through the cool, wet seasons and drying up as the weather turns warm. Might fish live in these types of ponds? Why or why not? Without fish in these ponds, how might survival be easier for other animals, such as tadpoles?
- Some ponds exist year-round. Do you think year-round ponds support more wildlife than temporary ponds?
- Look for animals that use the water as a source of travel above and below the surface. They might include ducks, fish, turtles, insects, tadpoles, and crawdads.
- Look for animals that take quick dips to the surface, such as dragonflies. How does animal life above the water compare to animal life below the water?

Rivers and Oceans

- How many different types of animals can you count along the shore?
- Are there tracks? Shells? Crustaceans?

Help Me Understand

Q: How do animals survive when they don't live near water, such as in a dry, hot desert?

A: Many desert animals get the moisture they need from the foods they eat, such as seeds and cactus fruit. Their bodies have adapted so they require very little fresh water to survive.

- Think about the surface of these large bodies of water and what other animals (birds, deer, bears, foxes, coyotes, and so on) might use them.
- Think about the many animals that live below the surface of these large bodies of water. How many can you list?
- Think of marine mammals that rely on large bodies of water. A marine mammal is an animal that spends most of its time on or under the water, even though it breathes air, has fur, and gives birth to live young (such as otters, seals, walrus, dolphins, and whales).

 Stimulates awareness of natural water sources

23

Stone Age

Rocks are as old as the hills and mountains. Each rock, even a tiny grain of sand, has its own story involving travel and time, a history that is unique to it. Just what was that grain of sand once? Where did it come from? Spend time outside with your child and think about the history that could have been a rock's life. How can a boulder become a grain of sand? The dirt you walk on was once part of rocks. Where did they come from? Who or what might have also walked on the very same rocks—a woolly mammoth, a saber-toothed tiger?

Rocks are made mostly of minerals. There are thousands of different kinds of minerals. Salt is one. Diamonds are another.

Get outside and search for different types of rocks. Walk slowly and carefully, selecting different shapes, colors, and textures. Examine their surfaces. Some rocks are hard, some are soft. Some rocks can crumble and be broken, others are solid.

Help Me Understand

Q: How do we know how old rocks are?

A: All rocks are as old as the Earth. Scientists who study rocks are called *geologists*. Studying rocks helps them learn about the history of Earth's societies, people, weather, and animals.

How do rocks change when they become wet? Place your rocks in water and watch what happens to them. Do their textures change? Do their appearances change?

Can you write or paint with rocks? Try rubbing your rock on a hard surface, such as a sidewalk. Can you carve with a rock? Try it on a soft surface, such as wood.

Ask your child to imagine how rocks might have been useful for cavepeople back in the Stone Age, two and a half million years ago. Explain that the Stone Age was given its name because rock was the principal source of tools back then. Today, people in most societies use metal for tools, but it hasn't always been so. Look at a rock, and let your imagination travel back in time.

☑ *Stimulates awareness and understanding of geology*

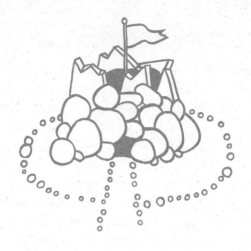

24

Rock 'n' Roll

When you walk, jump, or jog outside, chances are that you are stepping on rocks. Rocks make dirt when they're crumbled and mixed with organic materials. Small rocks were once part of larger rocks. Just what are rocks? Are all rocks the same? Get your rock hound outside to do some investigating about the hard facts surrounding these natural objects.

Take a walk with your child and search for rocks. Walk

slowly and look carefully. Make it a scavenger hunt and try to locate the following:

- A sharp rock
- A flat rock
- A bumpy rock
- A crumbly rock
- A rough rock
- A smooth rock
- A shiny rock
- A dull rock
- A rock with speckles
- A rock with stripes
- A multicolored rock
- A rock with only one color

Compare the different types of rocks you found. How are their textures different? How do they differ in appearance?

Play outdoors using stones and rocks to create an outdoor space. Show your child how small rocks can be lined up in rows to create a path or placed in a large circle or square to designate an outside space or a house. Create a town with roads, places, and spaces to visit, using rocks to designate each place you create. This activity works well with a group of children, as many hands make light work! Each child may create his or her own space using a line of rocks to designate it. Perhaps they'll build a rock fort with a secret password. If they create a town, encourage them to establish special businesses, such as an ice cream shop, a school, or a post office,

Help Me Understand

Q: Where do rocks come from?

A: Earth is a rocky planet, so the rocks in our environment come from many places, such as the ground, mountains, and volcanoes. Weather breaks rocks down over time, and they are carried to new places by rivers and oceans, gradually changing shape and form as they break or wear down.

within their rocky community. Allow imaginations to soar and grow as your young architects build and create while soaking in fresh air and sunshine.

Rock-Hard Facts

Rock is the main ingredient in earth. There are three types of rocks. Igneous rocks are the most common, sedimentary rocks are layered rocks (different materials pressed together over time), and metamorphic rocks have been heated so much that the minerals in them reform and make the rocks look wavy.

Examine the rocks you've collected and sort them as best as you can into one of the three categories: igneous, sedimentary, or metamorphic.

☑ *Stimulates exploration, curiosity, and observation skills*

25

Backyard Lullaby

It's easy to pop outdoors during the daytime with the sun to light our way to play. At sunset, however, most folks retreat back indoors. The warmth of summer offers a great opportunity to spend time outdoors at night with your child. There's a whole new world of nature to discover by night! Nocturnal animals come to life after a day of dozing. Stars become visible. The moon lights the sky as it reflects the sun.

Help Me Understand

Q: Why are some animals awake at night instead of during the day?

A: Animals that are awake at night are called nocturnal animals. Every animal occupies a certain place in nature, and this place depends on what it needs to survive. Nocturnal animals might come out at night to hunt for a certain type of food they like to eat, to avoid the heat of the day, or to avoid being hunted (because the darkness helps hide them from predators). Nocturnal animals have adapted to living the nightlife. Many, such as owls, have large eyes to help them see better in the dark. Others, like bats, use sound waves to help them "see."

Arrange a backyard campout with your child. Pitch a tent and spend the night sleeping outdoors in your backyard or on your patio. (If you can't spend all night outdoors, head out for a nighttime picnic or walk.) While you're outside, sit quietly and listen to the lullabies of nature.

What sounds do you hear? Frogs? Crickets? An owl or two? The scritch-scratching of some small creature as it forages for food? A coyote's howl? Discuss each and every sound you hear.

How do the sounds you hear outdoors at night differ from the sounds you hear inside at night? How do they differ from the sounds you hear outdoors during the day?

What natural sounds do you hear other than animal sounds? Is there a breeze blowing?

Snuggle in for a good night's rest, letting Mother Nature sing you to sleep.

☑ *Stimulates relaxation, curiosity, and appreciation of the natural environment*

26

Moon Shadows

Select a night with a full moon or a near-full moon to take your child outdoors to play by moonlight. (Check your local newspaper for when the next full moon will occur.) There is only one full moon a month, except for every few years when there are two full moons in one month. When this happens, it is called a *blue moon.*

Sit outside with your child by the light of the moon and allow your eyes to adjust to the dark. Seek out objects. What can you see? How many objects can you view by moonlight? Can you see trees? Plants? Houses? Each other? Make a verbal list of everything visible. If you live near mountains or tall trees, enjoy watching the moon as it makes its way from behind the objects that block your view. As it rises, you might even see its glow before you see the actual moon. If clouds are present, ask your child to watch what happens each time a cloud passes over the moon. Is there a silver lining? Does the sky go from dark to light? Look for animal shapes in the clouds each time they pass over the moon.

Watch the moon as it travels across the night sky. Relax together, and let your imaginations travel to the moon and back. Make a date to sit outside together again for the next night or two, and note the difference in the time the moon rises and the difference in its shape with each passing day.

Imagine you are a nocturnal animal, just awakening after a daylong sleep. Where will you travel tonight, and what will you do? Share your adventures.

Stand facing each other in plain view. How many steps backward can you take and still see each other? Can you play catch by moonlight? Give it a try.

What objects around you create shadows in the moonlight? Search for shadows of different shapes and sizes. Make your

Help Me Understand

Q: What makes the moon glow so brightly?
A: The moon does not give off its own light like the sun does. The light we see from the moon is actually sunlight reflecting off the moon's surface.

own moon shadows with your body. Allow the moon to be your night-light and play by its illumination.

Gaze at the moon. What details of its surface can you see? Are stars visible around the moon? Why or why not?

A Moon by Another Name

Most of us know the moon by several names—the harvest moon, the full moon, the new moon. But did you know there have been many names given to the moon throughout history? Native American tribes of the northern and eastern United States gave different names to the moon in order to track the seasons and time, such as full wolf moon (January, when wolf packs are known to howl in hunger), full worm moon (March, when the earth is moist from rains and worms surface to feed the birds, inviting spring), full flower moon (May, when flowers are in abundance), and full cold moon (December, when nights are long and cold), among others.

Celebrate your own full moon each month by recognizing something important in your family's life at that time and giving the moon a name that corresponds to that event. It might be birthday moon, in honor of your child's birthday; pumpkin moon, in honor of the pumpkins you harvested in October; or lost-tooth moon, to commemorate the month your child first lost a tooth.

☑ *Stimulates exploration and curiosity*

27

Twinkle, Twinkle, Little Star

Stargazing is a wonderful way to relax and let any tensions from the day slip far, far away. Provide your child with an opportunity to imagine and wonder infinitely as she gazes at the night sky, her thoughts lost in space. Sit outside at night with your child, look upward, and let the moon and stars do the rest. (In areas with light pollution from a city or the moon's reflection of light, it will be more difficult to see stars.)

Invite the shining star in your life to join you outside to watch the sunset, and while you relax together and watch the daylight slip into night, hunt for the first visible star in the darkening sky. The first person to spy a star gets to make a wish! Find the second visible star and wish on it too. Find the third, and the fourth, and the fifth visible stars. Make a wish for every star you see.

Help Me Understand

Q: How do scientists know how far away stars are?

A: Astronomers measure the distance of stars using mathematical formulas. The formulas are based on the position of a star in relationship to Earth and other stars.

As the sky darkens, continue stargazing with your child. Do all stars look the same? How are they different? (Some are brighter than others. Some look white or yellow. Some even look red.)

Scientists who study the stars know that a star's color reflects its temperature. Blue stars are the hottest, whereas red stars are the coolest. Have a stellar scavenger hunt with your child as you count how many red stars you can find in the sky, and how many superbright white or blue stars there are.

Ask your child what she wonders about the stars. Focus and gaze. Consider how far away each star is.

Stars are enormous. Our sun, the star closest to Earth, is just an average-sized star. It's not the biggest star or the smallest. If the sun were hollow, one million balls the size of Earth could fit inside it.

If stars are so huge, why do they look so small? Try this experiment with your child: Find a small to medium-sized object, such as a rock, and hold it up for your child to see. Ask her to watch the object as you continue holding it up while walking away. Walk and walk, creating a distance between your child and you. The farther away you are, the smaller the object will look. It works the same way with stars. They appear small, even though they are truly enormous, because they are so far away.

☑ *Stimulates, observation skills, infinite wonder, and imagination*

Fall

Activities for Cool Days
and Cloudy Weather

28

Time Out

We spend far too much time indoors, especially during the school year. By day, we're penned up in offices or classrooms under a ceiling and within four walls. At the end of the day, we often retreat to our homes. Even most modes of transportation keep us cut off from the outside. Fall offers an ideal time to take advantage of the outdoors, when it's not too hot or too cold, doing the ordinary in an extraordinary fashion.

Help Me Understand

Q: Why are we eating (or reading) outside?
A: The fresh air is good for us and helps to keep us healthy.

Do the daily things you normally do, but do them outdoors whenever possible. Be a role model for your children, and encourage them to take pleasure in simply being outdoors. Here are some ideas to get you and your kids started:

- **Story time.** Throw a blanket on the ground and read to your children outdoors. Select a weekend or evening, or perhaps find a spot under a tree at night with your flashlight in tow. This practice promises to make story time memorable.
- **Mealtime.** Make breakfast, lunch, or dinner a picnic under the sun, moon, or stars. You needn't go far. A porch, patio, or backyard will do. Take your meals outside and allow nature to add its own flavor to your food.
- **Playtime.** Take the toys outdoors. Scooper trucks and dump trucks, dolls, coloring books, whatever your children enjoy playing with. Move it all outside and give your children the opportunity to play with their favorite things in the fresh air.
- **In the car.** Drive with your windows open and feel the wind whip your face.
- **Hopscotch.** Chalk is inexpensive and fun. If you don't have access to a surface to draw on, scratch a hopscotch grid into a patch of dirt.

• **Hula hoops and jump ropes.** Let your children explore their imaginations while rolling, running, and jumping with a simple piece of rope or plastic circle.

Walk or ride bikes to get to places instead of driving, hopping a cab, or taking a bus. Once your family gets into the routine of walking and biking, think of other simple ways to do more things outdoors. Every little bit of effort to conserve the Earth's natural resources makes a big difference to the planet's health as well as to your family's!

☑ *Encourages exercise and well-being*

29

Zoom In

When we step outdoors with our children, it's easy to take nature for granted. We may walk through nature every day and never notice it. We are fortunate to be surrounded by infinite skies; clouds of amazing shapes, sizes, and colors; plant life; and a sun that provides constant energy. When we step outdoors, we walk on rocks, soil, and sand. We're surrounded by wildlife, even if we can't see it. Our ears give witness to

Help Me Understand

Q: Do all tiny spaces have animal or plant life?
A: Most spaces, even spaces we can't see because they're so small, contain some form of life.

what we can't see: branches blowing in the wind, bird chatter, cricket songs, or a coyote's call in the dark of night. Nature is all around us.

Explore nature up close with your child. Really close. Start with the obvious, then slowly, carefully seek out the nonobvious. First help your child become aware of the nature we see every day: trees, clouds, the sun, the sky. What else lives and exists outdoors? Discuss these wonderful and obvious gifts from nature with your child.

Now for the not-so-obvious. For this activity, select a spot the size of your child's foot in nature, and seek out animal and plant life within this tiny realm. You might select a spot in a field, a space on a tree trunk, a patch of lawn, a crack in the sidewalk, or the area under a rock.

Once you've selected your spot, discover what lives there. Look closely. Separate blades of grass and peek. Look as closely as you can. What can you find? Itty-bitty bugs? Fragile root systems? Discuss what you discover, and compare your findings to what you see at different places you observe up close.

Look carefully at other living and nonliving objects in nature near your home. You can do this anywhere, whether at a park or in your backyard. Use a magnifying glass to take in the intricate details and patterns found in a simple flower petal or leaf. Take in the swirls or twirls on a piece of bark, or the tiny specks that make up a grain of sand.

☑ *Stimulates curiosity and exploration*

30

A Thinking Place

Everyone needs to have some downtime in a quiet place where they're not interrupted—even kids. Or *especially* kids. It's healthy to provide your child with her own space to simply reflect, think, imagine, wonder, and be alone. Quiet time, without television, video games, or other stimulation, promotes thinking and relaxation, and it allows children to sort through their thoughts and feel calm.

Find the ideal outdoor thinking place with your child, and designate it as such. You might give it your own special name: the Quiet Spot, the Thinking Space, My Own Place.

Maybe it's at the base of an old tree, a spot on a porch that offers a view of the sky, or next to a window where you can be inside and look out. Spend time testing out and selecting your special space, just as you would spend time selecting other things that are important to you. Practice sitting in it and thinking in it. Practice being with it.

Help Me Understand

Q: Does anyone else use my thinking space?
A: Chances are, when you're not in your special spot, others visit it from time to time. We share the Earth with lots of living things.

Once your child has selected the perfect spot in nature, one that's serene and safe, allow her to use it as she sees fit. Give her the gift of time to simply sit—without interruptions and noise—and enjoy solitude. Place a chair or bench there, or provide access to a blanket or towel. Let your child know that the space is hers to use whenever she feels the need.

☑ *Encourages relaxation and comfort*

31

Wild Hide-and-Seek

It's hard to miss the sight of a brilliant cardinal splashing through the green of a tree's leaves. The red of its feathers provides such contrast. Animals are given specific markings and colors for a purpose. Male birds often have brilliant colors to help attract females. Sometimes people wonder why female birds get the short end of the stick in the appearance department, with their dull plumage. Consider this: the dull coloring helps mother birds blend into their surroundings while sitting on a nest incubating eggs, thus protecting them from potential predators. Aha! It makes perfect sense when the rules of nature apply.

Camouflage is nature's way of helping animals blend into their environment, an environment where they must constantly be on the alert as they search for food and try not to become food for another animal. It's a matter of survival.

Ask your child to look at animals in nature and think about how their markings and colors might help them survive in the

Help Me Understand

Q: How do animals get the markings they have?
A: Some markings are genetic, meaning they are inherited from the animal's parents. Some markings occur through adaptation—when an animal changes over time to be better suited to its environment—to help ensure the animal or plant's survival.

wild. A killer whale in the ocean is striking with black and white markings. The black on its back helps to camouflage it against the ocean's surface when seen from above. The white of its belly helps to camouflage it against sunlight hitting the ocean's surface when seen from below. Why is a polar bear white? Why is a rattlesnake brown?

Play a game of "wild" hide-and-seek with a group of children outside. Wear natural colors that will allow you to blend into the landscape. Start by tying a bright bandana to the arm of each child who will hide. Play by wild rules: rather than hiding completely from view, hiders must instead try to blend into their surroundings. Play again without the bright bandanas. How did the two games differ?

Observe animals in your area with your child, and discuss their markings and colors, and how each is "suited" for survival.

☑ *Stimulates sensitivity to animal life, observation skills, appreciation, and understanding of how animals survive in the wild*

32

Take a Walk
on the Wild Side

Everything in nature moves. Rocks shift. The Earth spins. Winds blow. Water travels. Animals hop, waddle, scurry, scamper, run, jump, crawl, and slither. The outdoors offers a wide assortment of movements, from a twirling leaf on its descent to the ground to an inchworm measuring its way upward on a flower stem.

Get your children outside and practice taking a walk on the wild side by pantomiming movements in nature. Play

Help Me Understand

Q: Does everything on Earth move?

A: Even though it may seem like certain objects in nature remain still at all times, everything on our planet moves. The Earth's surface is always gradually shifting, little by little, so even mountains move on occasion. Trees may be rooted, but their branches sway in the wind.

wild-style Simon Says. First select a leader. The leader will give commands to the game players, who must move toward the designated finish line—or the leader—via the movements the leader calls. For example:

- Simon says, "Waddle like a penguin."
- Simon says, "Twirl like a leaf."
- Simon says, "Slither like a snake."
- Simon says, "Stomp like a bear."
- Simon says, "Hop like a rabbit."
- Simon says, "Roll like a wave."
- Simon says, "Creep slowly like a sloth."

Create variations of the game by using specific types of movements, such as only using the actions of slow animals for one round, then using only those of fast animals for the next. Play again using movements in nature that aren't animal related, mimicking the wind, water, rain, waves, river currents, rolling stones, and so on. The first person to make it to the finish line is the winner and gets to be the next leader and call out wild commands.

What other traditional outdoor games lend themselves to "wild" games? Have fun creating your own adaptations of personal favorites!

☑ *Encourages exercise, creativity, and fun*

33

Some Web!

Spiders are amazing architects. A spiderweb is most often created to catch prey. Though not all spiders create a web for this reason, many species do. A common type of web you might find is an orb web. An orb web resembles a bicycle wheel; it's circular with "spokes" radiating outward. Spiderwebs are made out of silk. All spiders produce silk, which may also be used to encase eggs in an egg sac or leave a trail called a *dragline* to help a spider keep its bearings.

Help Me Understand

Q: How do spiders use webs to catch food?
A: When an animal walks, crawls, or flies into a web, it becomes stuck. As it struggles to get free, the movements vibrate the web, alerting the spider that there's a meal waiting. The spider knows where to walk on the web without getting stuck and also may have a special oil on its legs that prevents it from sticking to its own web.

One need only seek out a web in order to see the intricate detail a spider is capable of producing. The morning is a great time to go exploring with your child and look for spidery silk in the form of webs, egg sacs, or even an anchor for an abandoned spider exoskeleton. (Spiders shed their skin as they grow, leaving the old one behind.) Just observe spiders and their webs, don't disturb them. And never try to hold a spider. Some species are venomous, and all spiders can bite!

How many different types of webs can you find? How are they alike? How are they different?

☑ *Stimulates observation skills and awareness of and curiosity about the living world*

34

Roly-Poly Races

Roly-polies, also known as pill bugs or wood lice, are those little armored critters that kids seem to adore. They are crustaceans, so they're related to the crab family. They also play an important part in keeping the soil healthy by eating organic, decomposed material. This helps keep the soil not only clean, but also rich in nutrients.

Because roly-polies are crustaceans, they rely on gills to breathe. They live on land in moist areas and can usually be found under logs, rocks, or leaves. They have seven pairs of legs and roll into a tight ball when they feel threatened.

Seek out the roly-polies in your area, and watch them in action. If the day is not too warm or dry, you can encourage your child to pick one up gently and observe it up close. Roly-polies don't bite, but their movement on your hand might tickle, so handle them carefully.

Before you place your roly-poly back in its habitat, try this fun activity. Draw a small circle in the dirt with your finger.

Help Me Understand

Q: Are roly-polies bugs?

A: Many people call them bugs, but true bugs have a mouthpiece that can pierce through something and suck, such as an aphid's mouth, which will pierce a plant leaf and suck the juices out. So technically, the roly-poly is not a true bug. It's a land crustacean. There are many types of crustaceans. Most live in or near water, such as crabs, shrimp, and lobsters.

Place two roly-polies in the circle and watch them go. The first one out of the circle wins the race.

In your search for roly-polies, did you come across any other crawly critters? Beetles? Worms? Ants? Slugs? Snails? Take note of which insects and bugs, such as worms and roly-polies, are found in moist places.

Search outdoors for other types of bugs. Where are they found? Search on tree trunks, in grasses, and along sidewalks. Count how many different types of bugs you can find. If you live where fireflies come out at night, seek their flashes and enjoy nature's light show.

☑ *Stimulates curiosity and awareness of different types of animal life*

35

Look Up, Look Down, Look All Around

All kids love to spy and play detective. Move the mystery and enthusiasm outdoors and encourage your children to spy on the wildlife that makes its home near you. Lots of animals coexist with people, from wee spiders to grazing deer, depending on where one lives. It's simply a matter of uncovering the evidence of wildlife that lives nearby. Time to sleuth!

Help Me Understand

Q: Where do animals go during the day?
A: Even though we might find evidence of animals without seeing the actual animal, they're more than likely still nearby. A burrow's entrance might house a mouse that is sleeping by day. A cracked nutshell may have been dropped by a bird who was having lunch and then flew off in search of more food to eat.

Encourage your child to look for these signs of wildlife:

• A spiderweb
• A slimy snail's trail
• A chewed or nibbled leaf
• A track in the dirt, snow, or sand
• A feather
• A snakeskin
• A nest
• A burrow entrance
• A cracked nutshell

Look for other signs that animals share your environment. Can you find blooming flowers and grasses? Without pollinators, such as bees, butterflies, and bats, flowers wouldn't exist.

What other evidence can you find that animals, tiny or large, coexist with you? Gently look under rocks or fallen logs to discover what might be living beneath. Back at your home, look closely as well. Can you find evidence of animal life on or around your home? What has been there? Be a detective, and discover wildlife at work.

☑ *Stimulates awareness of one's surroundings*

36

Cloud Racing

What person hasn't stopped to gaze at clouds, daydreaming and getting lost in imagination? As an adult, you probably have fond memories of cloud watching as a child. Will your children share those same memories? This quiet time helps stimulate imagination and brain function. Provide your child with the opportunity to simply sit, think, and imagine. You will be giving him the treasured gift of time.

   Spread out a blanket or towel and lay down so you can stare at the clouds. Make certain the sun is not shining directly

Help Me Understand

Q: What are clouds?

A: Clouds are made of water droplets and ice. They're white because light from the sun shines and reflects on the water, all the way through the cloud. When they're gray, it's because they're so thick with water that light can't get inside them and reflect.

in anyone's eyes. Watch the clouds, then watch them some more. Ask your child to imagine that he's traveling on a cloud. What does he see? Where is he going? Have him make up a story about his cloud travels.

What shapes can you find in the clouds? Are they thin and wispy? Fluffy and puffy? Are they the same shapes as the clouds you will see tomorrow? Why or why not? If it's a breezy day, have a cloud race with your child. Each of you gets to select a cloud as your own.

Designate a landmark, such as a treetop, in the distance. The first cloud to reach the landmark is the winner of this race. Watch as they drift, until a winner reaches the mark.

☑ *Stimulates imagination, curiosity, and relaxation*

37

Wandering Wind

Hair-blowing, hat-tossing, whistling wind. Your child can't see it, but he can feel it. And we can see wind in action as it moves against things. Get your child moving just like the wind by exploring and experimenting with it outdoors.

There are many different types of wind to play with. Stand outside and determine what type is present:

- A calm wind—Everything around you is still.
- A slight breeze—There is a faint rustling of plants or leaves.
- A strong breeze—Large branches and treetops sway.
- A gale-force wind—It's difficult to walk against the wind, and tree branches start snapping off. (Get back inside! The wind is too wild to play with today.)

While outdoors, watch the wind in action. In which direction do you see the leaves blowing, grasses swaying, or bodies

Help Me Understand

Q: Just what is wind, anyway?
A: Wind is air. We normally don't feel air unless it's moving. When it moves, it's called wind.

of water rippling? Wind travels all over the world, moving air from place to place. Winds that blow from the south may bring warmer weather. Winds that blow from the north may bring cooler weather. Observe what objects move in the wind. Can you tell where the wind you feel might be coming from? Is it cool or warm?

Play a game with your child, and imagine where the wind has come from, what it has passed over, traveled across and through. Take turns imagining the faces and places the wild wind has touched.

☑ *Stimulates imagination, awareness of weather patterns, and curiosity about the environment*

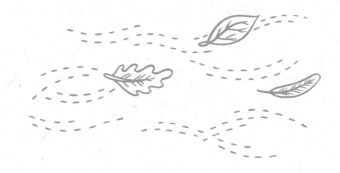

38

Say, "Whoosh!"

Your child is now aware that wind is air in motion. She has felt the wind and watched it in action. Now it's time to think of wind as a force in nature. Let's consider the wind as an energy source.

Stand outside together and listen to the wind. How many different sounds do you hear that the wind is responsible for? Are leaves making music? Are objects being moved by the wind, bumping against other objects, making noise as they do so? Can you hear water lapping against a dock? Chimes

Help Me Understand

Q: Why should we care about wind?
A: Wind is used as a power source. For example, windmills pump water. Sails on boats move them across the water.

tapping one another? How many sounds can you hear? Listen carefully. This is wind in action as a force in nature.

Play a game that shows the force of the wind. Gather different natural objects: a seed, a blade of grass, leaves of different sizes and shapes, small pebbles, rocks. Take turns throwing objects into the wind or blowing on or fanning the objects, and see how far you can make them move. Which objects move easily? Which objects won't move at all?

Have a wind race. Each of you should select one object found in nature. Establish a start line and a finish line on a smooth surface outside—this can be a crack in the sidewalk or a line drawn in the dirt.

Place your objects at the start line, then make wind. Race to the finish line by blowing your object into motion. The first object across the finish line wins!

Experiment with racing with different types of objects. Which objects were easiest to move? Did the racing surface affect how quickly an object moved across it?

Fly a Kite

Kites are an inexpensive and fun way to play in and with the wind. Many different countries and cultures have created different types of kites over the years. Although kites may vary in shape, design, and size, each works in the same way. Because kites are heavier than air, they must fly by force. Can you think of other objects that fly through the air with force?

☑ *Stimulates creative play and problem solving*

39

What Goes Up Must Come Down

Spilled milk. Falling leaves. Rain. Waterfalls. Ocean waves. Everything on Earth is pulled by a force called *gravity*. Gravity is present in our entire universe, not just on our planet. Sir Isaac Newton (1642–1727) developed the concept of gravity after observing an apple falling from a tree. Explore the concept of gravity with your child and discover the wonder of its force.

Jump upward. What happens? (You land back on the ground.) Toss a stick or a ball up in the air. What happens? (It falls down.) The Earth's gravity pulls everything near its surface down.

Experiment with tossing different objects found in nature up in the air and observe what happens to them. Do some objects fall faster than others? Practice jumping and see how high you can go.

Help Me Understand

Q: Why does gravity pull things down?
A: Gravity is an attraction, a pull, between two objects. The larger, denser, and heavier the object, the more pull it has. Earth is the largest mass near us, so its gravity has the largest effect on us and everything around us, even the moon. That's why things fall to the ground.

Next, have a race with gravity. Select natural objects of any size, shape, or weight for the race. Explain to your child that each of you will drop your selected objects from the same height to see which one lands first. Select a starting height, something easily visible and measurable. Drop your items on the count of three. The first object to hit the ground wins!

What did you discover? Was it a tie? Regardless of weight or size, objects dropped from the same height will fall at the same speed.

Objects of different weights, sizes, and shapes always land at the same time if dropped from the same height, unless an object has a large surface area that interacts with air. Air gets in the way when leaves and feathers drop, causing them to drift a bit and slowing down their fall.

Experiment with this concept:

- Compare dropping two rocks of equal size from the same height. (They'll land at the same time.)
- Compare dropping two rocks of different sizes and different weights from the same height. (They'll land at the same time.)

- Compare dropping a light twig and a heavy rock from the same height. (They'll land at the same time.)
- Compare dropping a leaf/feather and a rock from the same height. (The leaf or feather will take longer to land.)

☑ *Stimulates wonder and curiosity*

40

Groovy Gravity

Gravity pulls, no doubt about it. Sit outside with your child and listen for gravity at work. Can you hear a pinecone drop? A wave crash on the shore? A raindrop splatter?

Use your senses to experiment with gravity. Together, collect items from nature: pebbles, twigs, blades of grass, sand, soil, water, a feather, a rock. Drop each item, individually, into the palm of your child's hand. Ask her how it feels when gravity pulls each object to land in her hand.

Do some items land more softly than others? (Lighter objects will have less impact when they fall.)

What happens when those same objects are dropped from different heights? Do they feel different? (An object dropped from a greater height will have a stronger impact than one dropped from a short distance.)

Smooth out some dirt or sand. Ask your child to collect some natural objects (such as rocks, pebbles, shells). Drop a

Help Me Understand

Q: Why does an object dropped from a high height land with a heavier impact than when it's dropped from a lower height?

A: Falling objects accelerate, or move faster, as they fall. This causes them to land with more force.

rock into the smooth soil or sand. What kind of impact did it make?

Now try dropping a smaller rock from the same height into the soil or sand. Does the size of the impression change?

Experiment by dropping the same object(s) from different heights, smoothing out the sand or soil in between. Do the falling objects make different shaped and sized impressions?

What happens if a rock is thrown into the soil or sand from an angle? What kind of impression is created?

 *Stimulates a sense of experimentation and discovery*

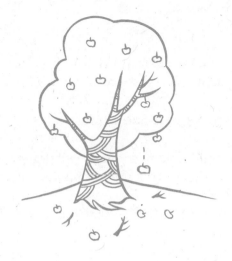

41

Falling for Nature

The gravity created by Earth pulls everything down to the ground. Search for objects in nature that have clearly been affected by gravity. Fall is the perfect time to seek out gravity in action, as leaves begin their tumble downward from the trees.

Look around. What can you find that has been pulled to the ground? What objects (leaves, nuts, twigs, rain) were once in

Help Me Understand

Q: What would happen if there were no gravity?
A: Everything not secured to Earth would float away.

a higher place but have fallen? Make a list of everything you can find.

Ask your child what might happen if gravity didn't exist on Earth. Here are some cool things to think about:

- The gravity from Earth keeps the moon from flying off into space.
- The gravity from the sun keeps the Earth in orbit around it.
- The mass of the moon is less than the mass of the Earth. Therefore, the moon's gravity is six times less than the Earth's. This means the moon has less gravitational pull than the Earth. If an elephant weighs 12,000 pounds on Earth, that same elephant would weigh 2,000 pounds on the moon. How much would you weigh on the moon (divide your weight by six)?
- An object, such as a rock, dropped on the moon would fall with less force than when it is dropped on Earth.
- Air affects how things fall. Earth has air. The moon has no air. Ask your child how he thinks this might affect falling objects in each place.

Go on a scavenger hunt and collect different types of leaves that have fallen from trees. Compare your treasures. How many different leaf shapes did you find? How many different colors? How many different textures?

☑ *Stimulates questioning and wonder*

Winter

Activities for Cold Days
and Snowy Weather

42

Tweet Treat

What if you had to stay outdoors all day and night in freezing temperatures? On top of that, imagine that your main food sources disappear or freeze during this time too. And the water source you might normally drink from may freeze. Winter is a challenging time for many animals. It is also a very quiet time outdoors, because many animals have adapted to winter by seeking shelter where they hibernate or by migrating to warmer climates. You certainly will not hear the chirps and buzzing of insects in the wintertime as you would during the summer. However, if you go outdoors during the winter, you may definitely still hear certain birds in action, because many species do not migrate to warmer places. As long as they have food sources and shelter, certain bird species can stay warm enough through the winter to survive. They remain in your region year-round, regardless of weather.

Give the birds in your neighborhood the gift of some delicious treats. Explain to your child that birds become very hungry during the winter months and often have difficulty finding food to eat. Place food out, such as orange wedges, seeds, and nuts. Get creative and make your own bird-food recipes by mixing together natural ingredients that birds love, such as raisins; peanuts (unsalted); sunflower seeds (unsalted); and peanut butter mixed with cornmeal, raisins, oatmeal, and seeds.

What do you do with the food? Peanut butter mixtures can be smoothed into pinecones and hung from tree branches or bushes. Orange slices and wedges can be skewered on sturdy

Help Me Understand

Q: Why do some animals stay where it's cold all winter, without hibernating or migrating like other animals do?
A: Animals that survive winter weather without hibernating or migrating have adapted to their environment. Adaptation means fitting in to survive. They may be able to insulate their bodies well by gaining weight before winter or by fluffing up their feathers to keep warm. They may have enough food sources available to help them stay active through winter weather.

sticks like shish kebab and placed in the junctions of tree branches.

Once your snacks are in place, it will be just a short time before the "bird word" gets out, and soon you'll have flocks of feathered friends visiting.

If there is enough snow on the ground, build a snowman feeder for the birds. Sculpt a snowman. Place a branch on each side of the snowman for arms, as you normally would. Dangle treats from the branches, such as peanut butter–covered pinecones dipped in birdseed. Plant a clay water dish on top of your snowman's head and fill it with birdseed, nuts, and dried fruits and berries. Before placing a carrot nose in your snowman's face, attach orange slices along it like a shish kebab. Use pine boughs or dried branches as the hair, sticking them on each side of the snowman's head, so birds will have a place to perch while awaiting their turn to dine.

It won't be long before your snowman becomes a very popular person! As the weather permits, replace the bird food as needed.

☑ *Stimulates caring and stewardship for living things*

43

Snow Me Some Fun

Snow days are the perfect time to bundle up and get outdoors for some fresh air and chilly, silly fun. Layer and bundle your kids and then propose these activities.

Build an obstacle course in the snow. Use the snow to build low walls, create tunnels, form arches, create pathways, and sculpt low landmarks or poles. Once your icy architecture is in place, have a race, encouraging your kids to jump, slide, glide, run, or hop to the best of their abilities over, under, around, and through the obstacles created in the snow. Use a stopwatch to gauge the time it takes each child to maneuver around the course.

Here are some variations of chilly, silly snow games:

- Build a ramp or small hill with snow and explore what materials slide more easily than others. Do plastic items slide more easily than those made of fabric? What type of

Help Me Understand

Q: Why does snow stick together?

A: Not all snow sticks together. If it's too cold—below 28 degrees with the ground also below freezing or 32 degrees—the snow may be powdery and probably won't stick. But water is cohesive, meaning water sticks to water (and other surfaces). If the temperature is near freezing with the ground temperature above freezing, snow may be warm enough to build with because it will be wet and it will stick.

surface moves the most easily on snow? Will some natural objects slip and slide on the snow? Experiment.

• Play "freeze tag."

• Have a contest to see who can jump the farthest in the snow.

• Have a contest to see who can make the most snowballs in three minutes.

• Have a contest to see who can build the biggest snowball.

 Encourages exercise and creative play

44

Fluffy Flakes

Some people believe that the first snowfall of the season is the most special. There's something about the silent flakes making their way down from the heavens that transcends our thoughts, and it's easy to get lost in the magic of simply watching snow fall to the ground.

Bundle your wintry kids up and get outdoors to examine the wonder and art of nature in the form of snowflakes. Wear dark clothing (a dark coat, gloves, or scarf) if you can, or sim-

Help Me Understand

Q: How are snowflakes made?

A: The moisture in clouds must be 14 degrees or below for an ice crystal to form. Once the perfect temperature and moisture conditions are present, ice will form a hexagon-shaped (six-sided) prism. Each time the ice crystal moves through different temperatures, it changes form, sprouting branches from the six corners of the hexagon. Wilson A. Bentley (1865–1931) is the first known photographer of snowflakes. He became interested in snowflakes as a child, and many books have been written about him.

ply take an extra piece of dark cloth out with you. If the weather is warm, the snow may be clumped into huge flakes (which are actually many smaller individual flakes stuck together). The chillier the weather, the better the chance to catch and observe individual flakes.

If the atmosphere is moist and cold enough, ice crystals will form. Crystals are six-sided and develop in the atmosphere with amazing symmetry. A snow crystal is a single crystal of ice. A snowflake may be made up of several crystals stuck together. Many believe that no two ice crystals are identical, but they all share the same symmetry—six sides, often with intricate branches and designs.

Hold out your dark cloth, be it your arm, your hand, or a piece of material, and catch ice crystals. Observe them closely. If you breathe on them, they may melt, so be careful as you study them. Compare the crystals and flakes you catch, noting their beauty, the intricate detail, and the amazing assortment of designs that nature has created.

If you begin feeling winter's chill, dance around like a snowflake to increase your body temperature and warm up!

☑ *Stimulates observation skills, curiosity, and appreciation of art and science*

45

Cloud Breath

As the weather cools, we bundle up to spend time outdoors. In previous activities during warmer weather, we explored clouds and played in rain. Cooler temperatures offer the perfect opportunity to experiment with exactly how clouds are made.

On a chilly day, pile on the layers for some playtime outdoors. If there are clouds in the sky, make note of them. Moisture rises from the Earth, and as it rises, it cools. The cool droplets of water condense (stick together) to form clouds, because water likes to stick to water.

Help Me Understand

Q: Do people make the clouds we see in the sky?
A: No. The clouds we see in the sky are created by the forces of nature on Earth, such as moisture from the oceans and from different types of weather. Sometimes there are manmade clouds we can see, though, such as vapor trails left from airplanes.

Demonstrate for your child. Breathe out into the cold air. You should be able to see it. This is a cloud! It is warm, moist air, or water vapor, from your body suddenly hitting cool air, which condenses it and creates a cloud. Eventually, the cloud disappears as it floats away and dries up, dissipating.

 *Stimulates curiosity and exploration*

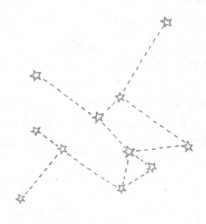

46

Dot-to-Dot Stars

A long time ago, people looked to the night sky and created pictures with stars. The stars were grouped as patterns or pictures known as *constellations*. Grouping and naming the stars helped people recognize specific areas in space and gave order to the night sky.

Wintertime is a great time to stargaze because the cold air is often dry, which offers better visibility. When you look out at the night sky again and again, you'll notice that the patterns the stars create are fairly constant. Knowing the placement of

Help Me Understand

Q: How many constellations are there?
A: There are eighty-eight constellations. We see different views of the universe depending on where we live. And with the Earth orbiting around the sun each year, different constellations always come into view.

stars in the sky helps give us a sense of where we are on Earth, like a map in the sky.

Spend time with your stargazer and look into the night sky. Obtain a constellation guide from your local library or visit NASA's web site (spaceplace.nasa.gov) and print out a constellation guide. How many constellations you can identify? Can you find the Big Dipper? Actually, the Big Dipper is not a constellation; it's part of the constellation Ursa Major, or Big Bear. It's easy to spot, because it looks like a giant spoon!

Encourage your child to look up into the sky and create his own patterns and shapes. Play dot-to-dot with the stars, and make your own designs. Create an animal shape, or find a square or triangle made from stars. Give your designs their own names.

☑ *Stimulates concentration, observation skills, and imagination*

47

Bare Naked Trees

The winter solstice is the official first day of winter in the Northern Hemisphere. It is the shortest day and longest night of the year and falls on or near December 21. Even though the winter solstice may mark the first day of winter, there are other signs we can look for in nature to tell us that winter has arrived: geese flying in V formation, heading south to warmer temperatures; cooler days; less sunlight; and bare trees.

Help Me Understand

Q: Why do trees lose their leaves?
A: There's less daylight in the winter, and leaves need sunlight to function and make food for trees. Because of this, deciduous trees (trees that lose their leaves in the fall and winter and grow new ones in the spring) drop their leaves as the days get cooler, and sprout new ones in the spring as the days warm up. During wintertime, deciduous trees "sleep," in a sense. Losing their leaves is part of their process for shutting down for the winter.

Observing bare trees in the winter affords a great opportunity to see trees in a way we might not see them during other seasons when they're full of leaves.

With your child, take note of the following:

- Tree shapes. Look at all those branches! Are the trees in your area shaped the same, or are there a variety of different types, shapes, and forms?
- How bare trees dress the horizon. Look at the sky through their bare branches.
- How the ground around winter trees appears so much brighter on sunny days, without leaves to create an umbrella shade.

Pick a bare tree and sketch it, making note of its many branches. Hold a scavenger hunt for nests. Now, with branches baring all, nests in trees pop out in all sorts of places:

nooks and crannies, up high in the tops, and lower near the trunks. How many nests can you spot? How many different types of nests do you see (squirrel nests, nests made of twigs, nests made of grasses)?

☑ *Stimulates awareness of the environment and concentration*

48

Snow Tales

When snow blankets the landscape, it seems as if everything becomes still in the world around us. But actually, animals and birds continue to forage for food, and all that's on the to-do lists of wintry wildlife becomes visible and evident, for a fresh blanket of snow often tells a tale.

Spend time exploring a landscape, be it a balcony, back-yard, playground, or field, that's covered with snow. Seek evidence of tracks and trails. Can you locate tracks? Are they from paws or claws? Where are the most tracks located? Out

Help Me Understand

Q: Snow will show tracks and trails. What else shows tracks and trails?

A: Lots of things! Sand, mud, dirt, water—explore and discover new ways to leave tracks.

in the open? Near a tree base? Around a woodpile, bushes, or shrubs? Along a balcony railing or snow-covered floor? A track is evidence of one animal having traveled over the snow.

If you have access to a rural or open space, can you locate a trail? A trail is a pathway used repeatedly by several animals. Where does the trail lead?

Put a bird feeder out for the wild birds to enjoy. After several hours, check to see how many different tracks you can find around it. Are there trails around it as well?

Create your own tracks in the snow with your feet, mittened hands, or a sled. Try making different types of tracks:

- Hop.
- Place one foot directly in front of the other to make a single line.
- Walk backward and forward; compare the two tracks.
- Walk sideways.
- Lie down and make a snow angel.

In what other ways can you make tracks and trails in the snow?

✓ *Encourages exercise and observation skills*

49

Feathers 'n' Fur

Imagine having to spend time outdoors in the winter—from sunup to sundown—with no jacket, scarf, or mittens. Imagine the animals who are active in the winter. They don't don extra clothing when the weather cools, but they have other ways of layering to keep warm. Just how do they do it?

Birds fluff their feathers to create extra insulation. Mammals grow thick fur coats to help insulate them. Foxes, squirrels, and coyotes may use their tails as blankets, wrapping them over their faces and noses.

Take a walk outdoors in chilly weather and search for a bird with its feathers fluffed.

When we get chilly, we can go back into our homes where there are walls to keep the chill out, heat, and maybe even a fireplace. Animals in the wild have to seek their own shelter from the cold. Where do they find it?

Look around with your child for evidence of animal homes. Can you spot nests in trees? Burrows in the ground? Twigs or

Help Me Understand

Q: What if an animal isn't able to fluff its feathers or grow thick fur to keep warm in the winter?

A: All animals have ways of adapting to seasons. Those who can't tolerate extreme cold may migrate to warmer places or burrow in shelters to stay out of the cold. Even some insects, such as bees and ants, huddle together in their nests in the winter.

leaves piled in the woods or thickets? What might be nesting there? (Insects or small animals.) Don't disturb any nesting sites you may discover. Simply observe.

When cleaning up leaves from your yard, consider keeping a small pile in an out-of-the-way place so animals or bugs can find shelter from the cold.

Place yarn scraps or small fibers and fabric scraps in the low nook of a tree or on a bush, so that animals can add them to their nests.

☑ *Stimulates compassion and stewardship for living things*

50

Slip, Slide, Wild Ride

When snow and ice cover the landscape, it can make getting around a little more challenging for everyone, wildlife and humans alike. It can also make traveling a lot more interesting and even fun.

Certain animals, such as rabbits, hop on snow, while some, such as deer and even people, walk on it. Animals such as lemmings tunnel beneath it. Some animals leap through it, plow through it, waddle on it, slip and slide on it, and even climb through it!

Help Me Understand

Q: What if the snow is too deep for a small animal to move through?

A: If the snow is deep, most tiny animals are light enough that they can move on top of it without sinking. Some animals have special physical characteristics to help them, such as the snowshoe hare, whose big feet act like snowshoes to keep it from sinking in deep snow.

Pantomime with your children, replicating how different animals might move through the snow. Seek out animals that are active during the day, such as squirrels, turkey, deer, and small birds, and watch them in the snow. If you're near a water source, observe ducks and geese as they maneuver on ice, in the water, and across snow.

 *Encourages exercise and creative play*

51

Wild and Wintry

Cold winters are certainly a bit quieter than the summertime, regarding what's out and about. It's a less active time for many species that adapt to cold temperatures by seeking shelter or migrating to warmer climates. However, it's not a completely vacant time in nature. Many animals remain active through-

Help Me Understand

Q: We found evidence of animals, but where are they?
A: Many rest and remain still so they can store up their energy, since finding food is more difficult now. They might be in a tree or a nest; within pine boughs; or in bushes, logs, or burrows.

out the winter, even in the coldest temperatures. You need only look and see.

Have your children search for animals that are present in the wintertime, such as cardinals, owls, deer, squirrels, blue jays, nuthatches, chickadees, titmice, bunnies, foxes, and so on. Even though you may not see an animal, chances are you'll see evidence that it has been around. The lack of foliage on trees makes spying a bit easier, however, and animals leave tracks and trails through the snow. Chances are you'll have great luck finding wintry wildlife.

Look for evidence with your children, be it in your backyard, at a park, or throughout your neighborhood:

- Food caches, such as seeds and nuts. Look but don't touch. Animals hid these food stores specifically to help them survive the winter, when less food is available.
- Chew marks. Many animals will nibble and eat bark from trees, since leaves are sparse. If you're near a natural water source, beavers are probably burrowed in their dens, but chances are you can find evidence of their existence from chewed branches and logs.

- Tunnels and burrows in the snow.
- Sounds. Can you hear birdcalls? Squirrel chatter? A coyote's howl?

Keep a journal of your discoveries, and use a sketchbook to render what you see.

☑ *Stimulates curiosity and awareness of wildlife*

52

Winter Garden

Together you and your child have experienced winter with wildlife and have observed firsthand what a challenge this season can be for many species, regardless of your winter climate. Temperatures drop. Food sources vanish. Water may freeze. Shelter is vital for survival. It's a tough time for wildlife! As winter winds down and melts into spring, it's the ideal time to begin planning ahead for next year by planting a winter garden. Planning and planting a winter garden is a fun way for you and your children to celebrate spring, while also looking ahead to create a winter haven for the wildlife that stays with you.

First, plan. Check with your local landscaper or in books from your library and research types of bushes and trees that are native to your region and that also produce winter food sources, such as berries. Planting them in the spring will provide a wonderful food source for wildlife in winters to come.

Help Me Understand

Q: What's a native plant?

A: Native plants are plants that are specific to certain areas. For example, the saguaro cactus is native to the Sonoran Desert. It grows very well there but wouldn't grow well somewhere where it snows or rains a lot. By planting native flowers, trees, and bushes in the places where they are meant to grow, we help to restore the planet to its original state and provide shelter and food for the animals that live in those specific regions.

Evergreen trees provide insulation and shelter for many species and create an aesthetic contrast to a winter landscape.

Low-lying shrubs and thick bushes will offer beauty to your yard in the spring and summer, and shelter and protection to animals in the winter. Again, native plants work best, because they require less maintenance to survive and native wildlife will benefit from them.

As spring turns to summer and fall, if there's lawn in your yard, leave a small patch of it long instead of clipping it back. This will become a haven for insects and small animals, such as mice and frogs, that need to hunker down through the winter and try to keep warm. Better yet, plant some native grasses in a section of your garden and let them grow naturally. Grasses insulate the earth, the same way a jacket insulates you.

If you live in a warm winter climate, prepare your garden for winter visitors who will spend the season with you, such as hummingbirds.

As fall leaves tumble down, allow a patch of ground to remain covered with leaves instead of raking all of them up. This will provide a safe, cozy haven for many insects.

When cold temperatures arrive again, you'll enjoy your winter garden in all its glory. Keep a journal of the animals who find a haven there, thanks to your efforts during the spring, summer, and fall.

☑ *Stimulates stewardship and appreciation of wildlife living in harmony with the environment*

Resources and
Recommended Reading

Books for Adults

Diekelmann, John, and Robert M. Schuster. *Natural Landscaping: Designing with Native Plant Communities.* Madison: University of Wisconsin Press, 2003.

Kress, Stephen W. *The Audubon Society Guide to Attracting Birds: Creating Natural Habitats for Properties Large and Small.* Ithaca, New York: Cornell University Press, 2006.

Louv, Richard. *Last Child in the Woods: Saving Our Children from Nature-Deficit Disorder.* Chapel Hill: Algonquin Books, 2006.

Mizejewski, David. *National Wildlife Federation Attracting Birds, Butterflies and Backyard Wildlife.* Upper Saddle River, New Jersey: Creative Homeowner, 2004.

National Audubon Society Field Guide to North American Birds: Eastern Region. New York: Knopf, 1994. (See also the field guide to birds of the Western region.)

National Audubon Society Field Guide to North American Insects and Spiders. New York: Knopf, 1980.

National Audubon Society Field Guide to North American Rocks and Minerals. New York: Knopf, 1979.

National Audubon Society Field Guide to North American Trees: Eastern Region. New York: Knopf, 1980. (See also the field guide to trees of the Western region.)

Sibley, David Allen. *Sibley Guide to Birds.* New York: Knopf, 2000.

———. *Sibley Guide to Bird Life and Behavior.* New York: Knopf, 2001.

Books for Children

Aliki. *Those Summers.* New York: HarperCollins, 1996.

Brenner, Barbara. *One Small Place in a Tree.* New York: HarperCollins, 2004.

Bunting, Eve. *Sunflower House.* New York: Voyager Books, 1999.

Elhert, Lois. *Planting a Rainbow.* New York: Voyager Books, 1992.

———. *Leaf Man.* San Diego: Harcourt Children's Books, 2005.

Heller, Ruth. *How to Hide a Butterfly and Other Insects.* New York: Grosset and Dunlap, 1992.

———. *The Reason for a Flower.* New York: Penguin Putnam Books for Young Readers, 1999.

Himmelman, John. *Frog in a Bog.* Boston: Charlesbridge, 2004.

Hutts-Aston, Dianna, and Sylvia Long. *An Egg Is Quiet*. San Francisco: Chronicle Books, 2006.

————. *A Seed Is Sleepy*. San Francisco: Chronicle Books, 2007.

Martin, Jacqueline Briggs. *Snowflake Bentley*. Boston: Houghton Mifflin, 1998.

McLerran, Alice, and Barbara Cooney. *Roxaboxen*. New York: HarperTrophy, 2004.

Rey, H. A. *Find the Constellations*. Boston: Houghton Mifflin, 1976.

Rockwell, Ann. *Bugs Are Insects*. New York: HarperTrophy, 2001.

Ward, Jennifer. *Forest Bright, Forest Night*. Nevada City, Calif.: Dawn Publications, 2005.

Weindensaul, Scott. *National Audubon Society First Field Guides: Birds*. New York: Scholastic, 1998.

Wright, Joan Richards, and Nancy Parker-Winslow. *Bugs*. New York: HarperTrophy, 1988.

Websites

National Audubon Society: www.audubon.org

National Wildlife Federation: www.nwf.org

National Wildlife Federation Green Hour: www.greenhour.org

The Nature Conservancy: www.nature.org

The Sierra Club: www.sierraclub.org

About the Author

Jennifer Ward is the author of numerous children's books, all of which present nature to kids. Her award-winning titles have been featured in magazines such as *Ranger Rick, Your Big Backyard, Learning,* and *Foreword,* and she has been interviewed and featured on national television and local radio. She is a regular speaker at conferences and schools across the country, where she instills the importance of literacy and the wonders waiting to be discovered in the natural world.

About the Illustrator

Susie Ghahremani is a graduate of the Rhode Island School of Design. Her artwork, which combines her love of nature, animals, music and patterns, has appeared in the *New York Times, Nickelodeon Magazine,* and *Martha Stewart Kids,* and has received illustration awards from American Illustration, the Alternative Pick, and Giant Robot.

Born and raised in Chicago, Susie now happily spends her time painting, drawing, crafting, and tending to her pet finches and cat in San Diego, California. Visit her on the web at boygirlparty.com.